ALPHA BOOKS
RENEWABLE ENERGY

Nicola Barber

Evans

Evans Brothers Limited

This book is based on FACING THE FUTURE RENEWABLE ENERGY by Alan Collinson, first published by Evans Brothers Limited in 1991, but the original text has been simplified.

Evans Brothers Limited
2A Portman Mansions
Chiltern street
London W1M 1LE

First published 1995

Printed in Hong Kong

ISBN 0 237 51525 3

Acknowlegements

Maps and diagrams: Jillian Luff of Bitman Graphics
Illustrations: Outline Illustration, Derby – Andrew Calvert, Andrew Cook, Andrew Staples
Design: Neil Sayer
Editor: Su Swallow
Language Advisor: Suzanne Tibertius
Production: Jenny Mulvanny

For permission to reproduce copyright material the author and publishers gratefully acknowledge the following:
Cover photographs: (top left) Sullivan and Rogers, Bruce Coleman Limited, (top right) Mark Bolton, ICCE, (bottom left) Shelia Terry, Science Photo Library, (bottom middle) Molyneux Photography, (bottom right) Intermediate Technology.
Title page (Solar panel heating systems for heating water, Alanya, Turkey) Sheila Terry, Science Photo Library **p7** Alan Collinson, **p8** Sekal, Zefa **p9** C.C. Lim, John Topham Picture Library **p10** E. Landschak, Zefa **p11** (top) Lynn Baker, Compix, (bottom) Peter Ryan, Science Photo Library **p12** Zefa **p13** Peter Menzel, Science Photo Library **p14** Robert Harding Picture Library **p15** Dr Peter Thiele, Zefa **p16** Andrew Hill, Hutchison Library **p17** (left) Trygve Bolstad, Panos Pictures, (right) Mark Boulton, ICCE **p18** (top) Hutchison Library, (middle) David Lomax, Robert Harding Picture Library, (bottom) Sullivan and Rogers, Bruce Coleman Limited **p20** (top) Molyneux Photography, (bottom) "Club Med T, Club Med's Sailing Cruiser" **p21** Peter Menzel, Science Photo Library, (inset) Ken Lucas, Planet Earth Pictures **p22** (top left) Bramaz, Zefa, (bottom left and centre) Martin Bond, Science Photo Library, (bottom right) John Lythgoe, Planet Earth Pictures **p23** Voigtmann, Zefa **p24** (top and bottom) Martin Bond, Science Photo Library **p25** Martin Bond, Science Photo Library **p26** J. Pfaff, Zefa **p27** Robert Harding Picture Library **p28** (top) Alan Collinson, (bottom) Martin Bond, Science Photo Library **p29** Jonathan Wright, Bruce Coleman Limited, (inset) Images Colour Library **p30** Intermediate Technology **p33** (left) Handford Croydon, Zefa, (middle) Robert Harding Picture Library, (right) Walter Rawlings, Robert Harding Picture Library **p34** (top) George Holton, Science Photo Library, (bottom) Simon Frazer, Science Photo Library **p35** Alan Collinson **p36** Alan Collinson **p37** (top) Architects and Energy Consultants, The ECD Partnership, (bottom) Sally Morgan, Ecoscene **p38** The ECD Partnership (main picture: the Hughes Home, Milton Keynes) **p39** (top left) Philips Lighting Ltd, (middle left) Juliet Highet, Hutchison Library, (middle and middle right) Rockwool Ltd, (bottom) The ECD Partnership **p40** (top) General Motors, (bottom) Paul Shambroom, Science Photo Library **p41** Jery Mason, Science Photo Library **p42** Chris Howes, Planet Earth Pictures **p43** (left and right) Sue Cunningham, David Austen, Tony Stone Worldwide.

Contents

Introduction

Every day we use energy to give us heat, light and power. Energy comes from burning fuel. There are many different kinds of fuel, but the main ones are coal, oil, gas and uranium. All these fuels come from the Earth's **crust**. Coal, oil and gas are called **fossil fuels**. This is because they come from the remains of animals that died thousands of years ago. There is only a certain amount of each kind of fossil fuel. Once we have used up all the fossil fuels, the energy they contain is gone for ever. Another problem with fossil fuels is that as we burn them for energy, they **pollute** the air with waste gases.

All around the world, scientists and engineers are looking for new ways to provide energy. They need to find sources of energy that will not run out, and will not pollute the air. So far they have found four answers. These are the energy from heat in rocks inside the Earth, from salty water, from sunlight and from the power of gravity. These sources can provide power without pollution. Rubbish can also be used to provide energy,

but when it is burned it can cause pollution. These sources are called renewable energy sources. 'Renewable' means that these sources do not run out, no matter how much is used.

You can see the main renewable energy sources on the chart on the opposite page.

This book looks at some of the ways that renewable energy sources are already used around the world. In the next 100 years, scientists will find more ways to make use of the energy from renewable sources.

Words in **dark** type are explained in boxes at the end of each section. There are also **See for yourself** boxes with questions and ideas about energy for you to think about.

---

crust — the Earth's crust is the thin shell of solid rock that covers our planet.
fossil fuels — coal, oil and gas are fossil fuels. They come from the remains of plants and animals trapped in rocks for thousands of years.
pollute — make dirty

SOURCES OF RENEWABLE ENERGY

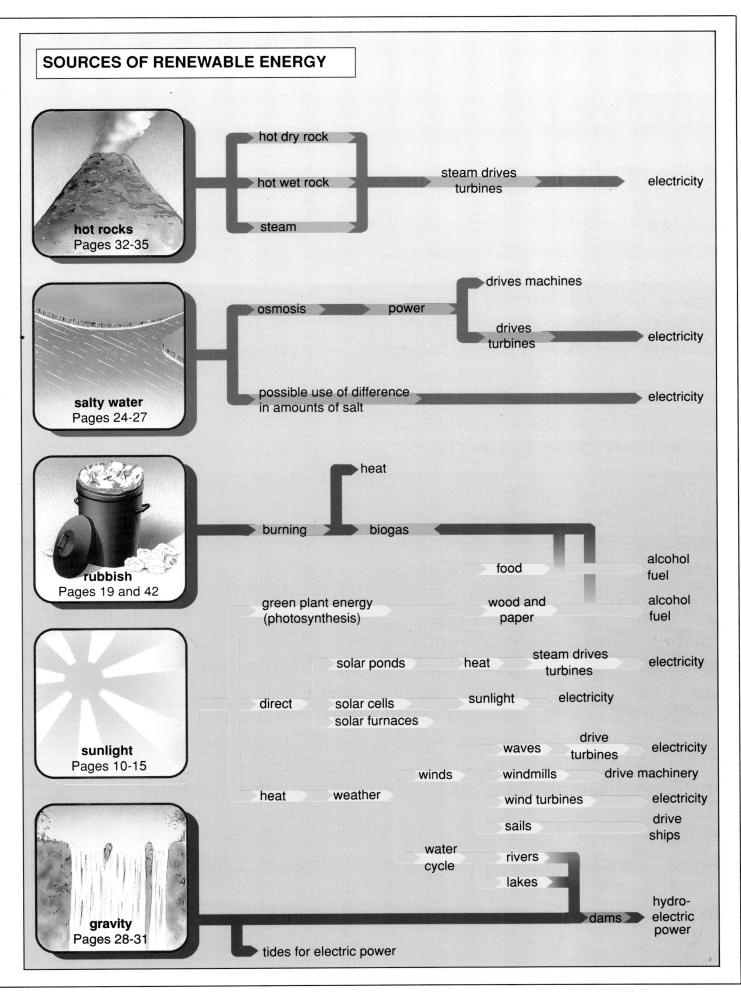

hot rocks
Pages 32-35

hot dry rock
hot wet rock
steam → steam drives turbines → electricity

salty water
Pages 24-27

osmosis → power → drives machines
drives turbines → electricity
possible use of difference in amounts of salt → electricity

rubbish
Pages 19 and 42

burning → heat
burning → biogas
food → alcohol fuel
green plant energy (photosynthesis) → wood and paper → alcohol fuel

sunlight
Pages 10-15

solar ponds → heat → steam drives turbines → electricity
direct → solar cells / solar furnaces → sunlight → electricity
heat → weather → winds → waves → drive turbines → electricity
windmills → drive machinery
wind turbines → electricity
sails → drive ships
water cycle → rivers
lakes

gravity
Pages 28-31

dams → hydro-electric power
tides for electric power

Energy in rich and poor countries

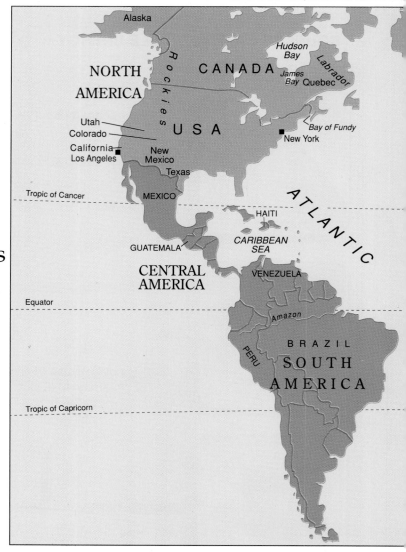

About 100 years ago, engineers invented the steam turbine. This was a large fan turned by a jet of steam. Engineers used the turning movement of the fan to make electricity. People used electricity to drive machines in new factories. These changes are known as the Industrial Revolution.

The Industrial Revolution happened in the countries of Europe, in North America, Australia, New Zealand and Japan. All these countries now use lots of energy to keep their factories, houses, offices and transport running. These rich countries are often called the developed countries (see map). But there are many other parts of the world where the Industrial Revolution did not happen. In most of Africa, South and Central America, India, China and Southeast Asia most people still make a living from farming. These poorer countries are often called the developing countries.

Rich and poor

Over half of the people in the world live in the poorer developing countries. But the developed countries use more than three quarters of all the world's energy. The poor countries would like to catch up with the rich, developed countries. But this is not easy because they cannot afford expensive power stations and machinery. In the future, it is likely that the developing countries will use cheaper, renewable sources of energy.

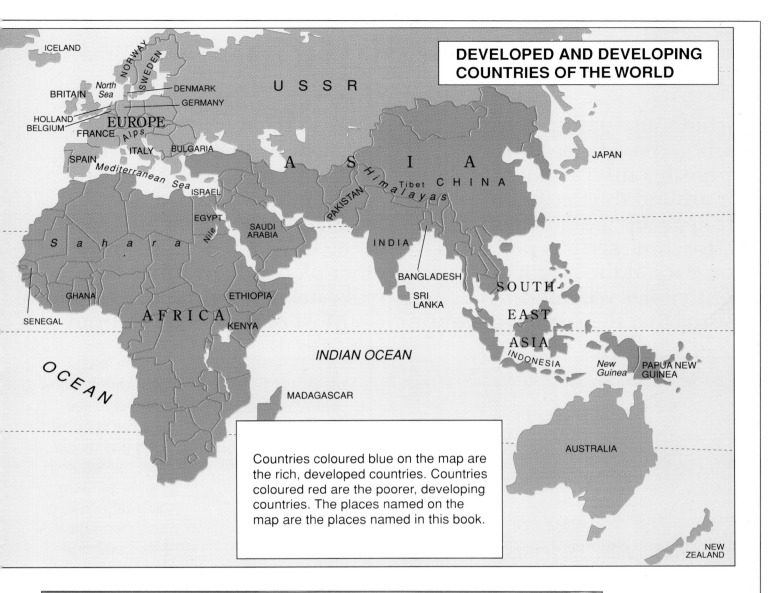

Countries coloured blue on the map are the rich, developed countries. Countries coloured red are the poorer, developing countries. The places named on the map are the places named in this book.

Fossil fuels cause pollution in the air and on the ground. This is waste from a coal mine in Durham, England. The waste is dumped on the beach. In 1993 the mines closed so the waste dumping stopped, but the beach will be spoilt for many years to come.

Energy and pollution

Greenhouse gases such as carbon dioxide (see box opposite) are warming up the air around the Earth. But if the air heats up too much it will cause some serious problems. Some ice at the North and South poles will melt. This means that there will be more water in the world's oceans and seas. So there will be danger from flooding in many parts of the world, such as Bangladesh, Holland and parts of the USA and Britain. In other places the warmer air will cause **droughts**.

Most of the pollution from

◄ One waste gas from fossil fuels is called sulphur dioxide. This gas mixes with the water in the air to make acid rain. When acid rain falls it kills trees and pollutes lakes and rivers.

► Smoke from **power stations** pollutes the air. The tall chimney on the right of the picture is giving off smoke from coal-burning, which is polluting the air. The other chimneys, called cooling towers, are giving off steam.

A 'greenhouse gas'

Burning fossil fuels pollutes the air. These fuels contain a chemical called carbon. When we burn these fuels, the carbon joins with another chemical in the air called oxygen. Together these two chemicals make carbon dioxide. The air around the Earth is warmed by the heat energy from the Sun. But carbon dioxide absorbs this heat energy. So the more carbon dioxide the air contains, the warmer it gets. The carbon dioxide is rather like the glass in a greenhouse. It lets the light and heat from the Sun in, but it does not let the heat out again. So people often call carbon dioxide a 'greenhouse gas'.

coal, oil and gas is caused by the rich developed countries. Many governments have already tried to cut down the burning of these fuels. But in the future, more energy must come from renewable sources that do not pollute the air.

See for yourself
Take two dry sticks and rub them together fast and hard. Now touch the place where the sticks touched. What do you notice? You have changed the energy of movement to heat energy.

power station — a place where fuel is burned to make electricity.
drought — a long period with little or no rain.

Energy from the Sun

The Sun provides the Earth with huge amounts of free light and heat energy. The heat energy from the Sun warms up the surface of the land and the sea. It also warms up the air around the Earth. People have always used the free heat provided by the Sun. If you look at the map below and compare it with the map on pages 6-7, you will see that many of the poorer countries of the world have lots of sunshine. This means that people can use a large amount of free heat energy in these countries.

These are greenhouses on a hill in Italy. The glass in the greenhouses lets sunlight through, but does not let the heat from the sunlight out again. The heat helps the plants in the greenhouses to grow (see page 14).

Reflected sunlight

People in hot countries can use the heat from sunlight to provide heat for

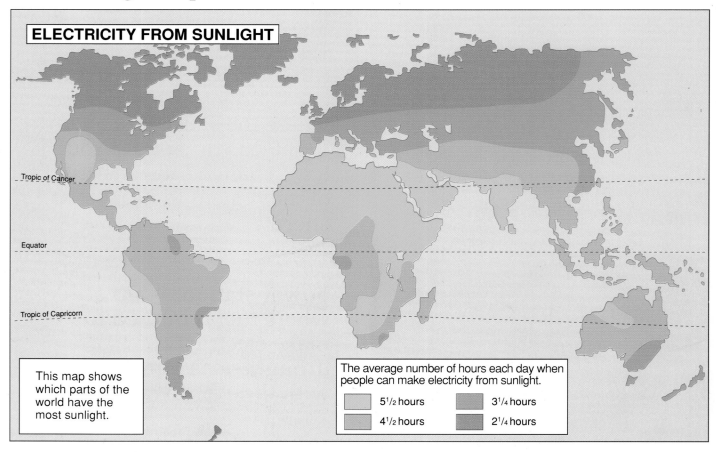

ELECTRICITY FROM SUNLIGHT

Tropic of Cancer

Equator

Tropic of Capricorn

This map shows which parts of the world have the most sunlight.

The average number of hours each day when people can make electricity from sunlight.

5½ hours

4½ hours

3¼ hours

2¼ hours

cooking. A cooker that uses heat energy from sunlight is called a **solar** cooker. A solar cooker has polished metal mirrors to reflect the sunlight. The mirrors concentrate all the heat energy from the sunlight on to one point. This concentrated heat energy is then used to cook food or boil water.

In some places, sunlight provides the energy to drive machines. Lots of mirrors are used to concentrate the heat energy from the Sun (see picture on page 13). This concentrated energy heats up oil inside a container called a solar furnace. The hot oil is then used to boil water. The steam from the boiling water drives turbines to make electricity (see page 6).

▲ ◀ Two kinds of solar cooker. In poor countries, people use heat energy from the Sun to cook food and boil water.

Water for drinking

In hot parts of the world, it is sometimes difficult to find enough clean drinking water. The water in the sea is too salty for people to drink. But in some places, people can use the heat energy from the Sun to make drinking water out of salty sea water. They do this by taking the salt out of the water. The energy in the sunlight is used to heat up the salty water. When the salty water is boiling, the water turns into steam. The salt is left behind. The salt is removed, then the steam cools and turns back into water again. The water is now pure and people can drink it. This process is called desalination which means 'salt removal'. The place where the salt is removed is called a desalination plant.

The Dead Sea, between Israel and Jordan, has very salty water. The salty water is made pure for drinking at a desalination plant (inset).

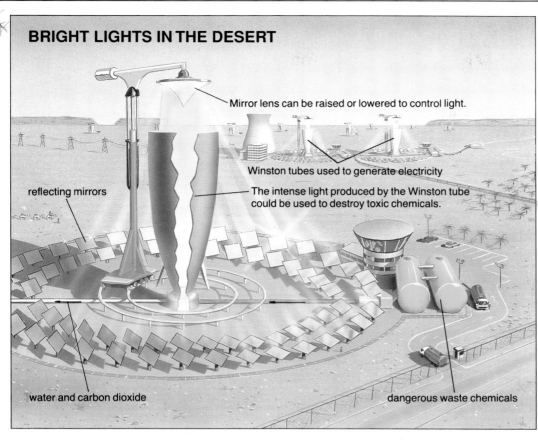

BRIGHT LIGHTS IN THE DESERT

Mirror lens can be raised or lowered to control light.

Winston tubes used to generate electricity

The intense light produced by the Winston tube could be used to destroy toxic chemicals.

reflecting mirrors

water and carbon dioxide

dangerous waste chemicals

◄ The Winston tube is a new invention. It concentrates the heat energy from the Sun. Winston tubes could provide energy to make electricity or to destroy polluting chemicals.

▼ This is a solar power station in California, USA. The mirrors reflect sunlight on to a tower. The heat heats water and steam from the water drives turbines to make electricity.

Space power and solar power

Sunlight can be made directly into electricity using a solar cell. Scientists first came up with the idea of solar cells to provide power for spacecraft in space. A solar cell contains chemicals which make electricity from sunlight. Solar cells are very useful in hot, sunny countries. For example, in Egypt electricity made from solar cells provides power for a main telephone line over 1000 kilometres long. The solar cells work because there is clear sunlight almost every day in Egypt.

In 1995 a new chemical was invented to make solar cells work even better. If the chemical is used in the future, many machines such as cars

Green plants use the energy in sunlight to make their food. This process is called photosynthesis. Will scientists be able to copy photosynthesis?

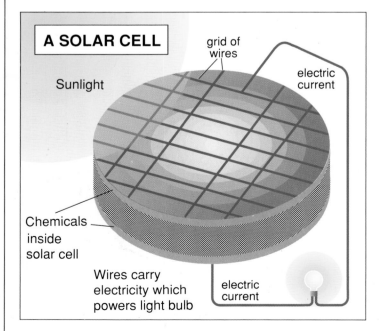

A SOLAR CELL

Sunlight

grid of wires

electric current

Chemicals inside solar cell

Wires carry electricity which powers light bulb

electric current

and ships will use solar energy. This will help to cut down pollution. And buildings will be able to make their own electricity.

Light and life

Sunlight is also used by green plants. Green plants take in light energy from the Sun to make food. The process of making food from sunlight is called photosynthesis. This means 'making with light'.

There is a special green chemical in the leaves of a

plant. This green chemical is called chlorophyll. When sunlight falls on a plant, the chlorophyll uses the energy in the sunlight to split water in the leaves into two other chemicals. These chemicals are oxygen and hydrogen. The plant does not need the oxygen, so it lets it out into the air. It uses the hydrogen to make food.

Scientists are now looking at ways of turning sunlight directly into chemicals. To do this, they want to copy the process of photosynthesis to get hydrogen out of water. Hydrogen is a kind of fuel. If there was a cheap way to get hydrogen out of water, it could replace coal, oil and gas as a fuel.

Some telephones get their power from the Sun. This telephone is in Australia. The panel above the phone box contains solar cells. The solar cells use the energy in sunlight to make electricity. The electricity powers the telephone.

See for yourself
On a sunny day take a small magnifying glass, a short length of cotton, a clear glass bottle and a nail or screw. Tie the nail to the cotton and suspend it inside the bottle. Focus the Sun's rays through the glass on to the cotton. Hold it very steady and see what happens after a while.

solar — describes something that is powered by energy from the Sun.

Green energy

Plants use photosynthesis to take in energy from sunlight (see page 14). Plants such as trees grow roots, trunks and branches by using this energy. When we burn wood from trees, this energy is given off as heat. So wood is a very important fuel in many parts of the world.

Sunderlal Bahugowa lives in India. He wants people to protect trees and plant more trees. He has already stopped people cutting down trees in many places.

People in many countries use wood to provide heat for cooking and heating their homes. Sometimes they use wood to power machines. But wood is not a very good fuel for most machines. Coal and oil are much better because they contain more concentrated energy than wood. However, coal and oil are not renewable energy sources. Wood is a renewable energy source because new trees are growing all the time.

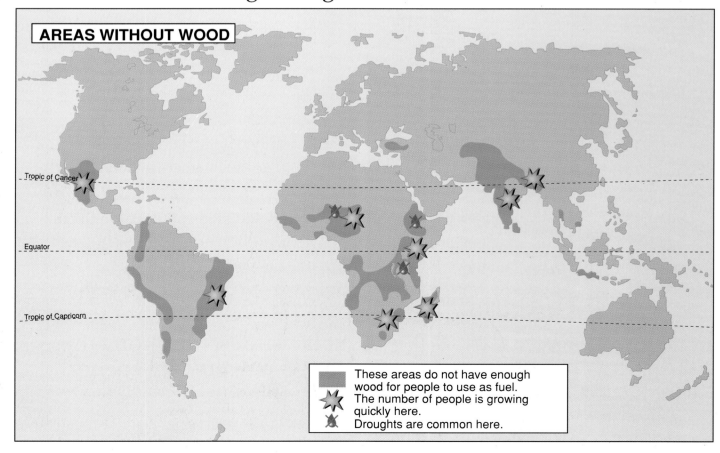

AREAS WITHOUT WOOD

Tropic of Cancer

Equator

Tropic of Capricorn

These areas do not have enough wood for people to use as fuel.
The number of people is growing quickly here.
Droughts are common here.

Collecting wood for fuel (above) and growing trees (left) in Kenya, Africa.

People use wood as a fuel. But that is not its only use. Wood is also used to make paper, some chemicals, furniture and buildings. As a result, in many parts of the world there is not much wood left.

How much wood is left?

The number of people in the world is growing very quickly. In many places, people have cut down more and more forest for fuel and for building. In poor parts of the world especially, there is not enough wood left. Another problem is that cutting down trees harms the land. The roots of a tree help to trap water, and protect the soil. When the tree has gone, rain can easily wash away the soil. If this happens it is often difficult for people to grow food.

Many countries have started to try to save their forests. They are also trying to plant new trees. But the trees will take a long time to grow. There will be a serious shortage of firewood by the end of this century.

Forests and pollution

Trees take in carbon dioxide as they grow. This means that they help to control the amount of

▼ The air in the city of Los Angeles, USA, is full of pollution from car engines. In the future, cars could run on a kind of fuel made from wood waste (bottom). This fuel would not pollute the air.

carbon dioxide in the air. If people do not plant new trees to replace the ones they are cutting down, there will be more carbon dioxide in the air. Carbon dioxide is a 'greenhouse gas' (see page 9). So there is a connection between trees and the pollution that is making the air around the Earth heat up. Connections like this are part of the science of **ecology**. Ecology is the study of how plants, animals and people are connected to their **environments**.

Wood-powered cars

People can use bits of waste wood to make fuels for vehicles. They use this wood to make two main fuels. These are both kinds of alcohol, called methanol and ethanol.

◀ Farmers grow sugar cane in Brazil. The sugar cane is used to make a new kind of fuel called ethanol.

▶ Biogas digesters are used in many countries to provide gas for cooking and lighting.

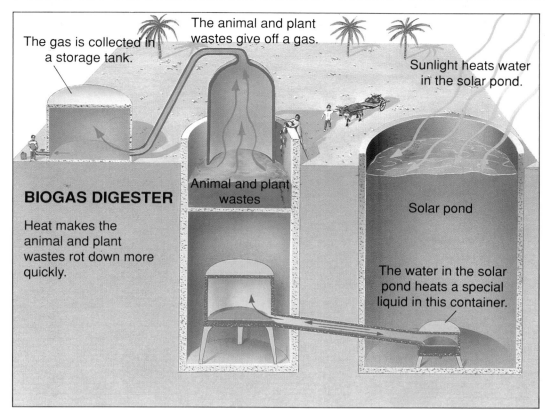

The gas is collected in a storage tank.

The animal and plant wastes give off a gas.

Sunlight heats water in the solar pond.

BIOGAS DIGESTER

Heat makes the animal and plant wastes rot down more quickly.

Animal and plant wastes

Solar pond

The water in the solar pond heats a special liquid in this container.

These two fuels cause much less pollution than the fuels made from oil (petrol and diesel) when they are burned in the engine of a car or truck. In Brazil, alcohol fuel is made from sugar cane. Many vehicles in Brazil use this new kind of fuel.

Energy from rubbish
Every day, people dump huge amounts of rubbish into holes in the ground. But soon this rubbish could be used to provide energy.

As the rubbish rots in the ground it begins to give off a gas. This gas is called methane gas. In some places in Great Britain, the gas is collected. It is then used to make electricity, and for central heating.

In some countries, people use a special machine to collect gas. This machine is called a biogas digester. Biogas is gas that comes from animal and plant wastes. In developing countries the biogas is used for cooking and lighting. Farmers in the Netherlands also use biogas digesters to get rid of the wastes from their herds of cows.

ecology — the science that studies how plants and animals are connected with their environment.
environment — the world all around us.

19

The energy of the wind

People have made use of wind power for years, to blow sailing ships along and to turn windmills. For the past 20 years, people have also used wind power to make electricity. On modern windmills, the wind turns flat blades called vanes. The turning movement of the vanes is used to make electricity. These modern windmills are called Wind Turbine Generators (WTGs).

There are about 40,000 WTGs in the world. They make about the same amount of electricity as three large coal-burning power stations. The wind is a free source of energy, and it will never run out. So why aren't there more WTGs in use around the world?

The answer is that fuels such as coal, oil and gas are cheaper than wind power. It takes a lot of WTGs to make quite a small amount of electricity. But small amounts of coal, oil and gas will provide the energy to make a lot of electricity. However, coal, oil and gas all cause pollution. It costs a lot to clean up the damage caused by the pollution. If governments start to include the cost of pollution in the price of coal, oil and gas, these fuels will not be so cheap.

▲ In some parts of the world, people still use old windmills to drive machinery.

Nuclear power? No thanks!

Some power stations use nuclear power to make electricity. Nuclear power comes from lumps of **radioactive rock** such as uranium. This is a kind of rock that gives

► Wind Turbine Generators (WTGs) in California, USA. The wind turns the vanes of the WTGs. A large collection of WTGs (inset) is called a wind farm.

New sails This large ship is called *Club Med 1*. It has engines but when the wind blows, the sails help to drive the ship. This saves fuel.

out a lot of energy in rays called radiation. The energy in radiation is so powerful that it is very harmful if it touches living things.

Many people thought that nuclear power was a cheap way to make electricity. A small piece of radioactive rock provides enough energy to make

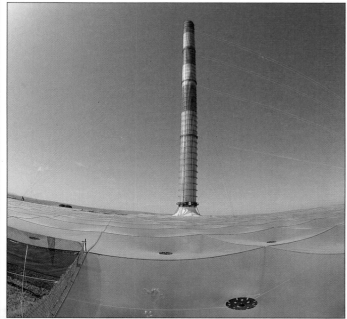

Designs for the future

No one knows yet what is the best shape for a WTG. Some WTGs have two blades. Some are round and some are bow-shaped. One design does not have any blades at all. It uses large plastic sheets which wave up and down in the wind.

a lot of electricity. But there is a problem when nuclear power stations begin to wear out. All nuclear power stations contain lots of radioactive materials. These materials give off dangerous radiation. So people have to be very careful with the materials in old nuclear power stations. This costs a lot of money, and makes nuclear power an expensive way to make electricity.

Some countries think that nuclear power is too expensive and too dangerous. Sweden has stopped building nuclear power stations. The USA has not built a new nuclear power station for over 15 years. But other countries, such as France, Germany and Japan, still make a lot of electricity from nuclear power.

Wind farms

In the future, people will use more WTGs to make electricity. It is important to put WTGs in

▶ An island in the Indian Ocean. In the future, WTGs may provide power for the people living on small islands such as this one.

the right place. In Britain, scientists use computers to test the direction of the wind, and to find where it blows most strongly.

On land or sea?
Many WTGs are usually put in one place. This is called a wind farm. Some people do not like wind farms because they think they spoil the look of the countryside. Other people say that the sound of the vanes turning in the wind is very noisy. In the future, more large wind farms will be built in the sea to move them away from places where people live.

radioactive — describes a material that gives off harmful rays of powerful energy called radiation.

Energy from the sea

A large part of the surface of the Earth is covered by sea. So far, scientists have found five possible ways of getting energy from the sea. They can use the energy of waves and tides. They can make energy from the salty water in the sea, and the heat in the sea. And they can use the currents in the seas and oceans. People already use the first four ways of making energy from the sea and sea water.

Testing a wave energy machine in Scotland

Wave power in Scotland

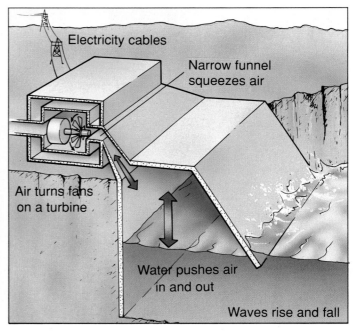

Electricity cables

Narrow funnel squeezes air

Air turns fans on a turbine

Water pushes air in and out

Waves rise and fall

As the waves break the water goes up and down. This movement pushes air in and out of a narrow funnel. The rush of air turns the fans of a turbine to make electricity.

This is the wave energy machine called CLAM. CLAM floats on the surface of the sea, riding up and down on the waves. This movement pushes air into the bags on the side. The air turns turbine fans inside the metal part in the middle of the machine. The turbines make electricity. A cable carries the electricity to the shore.

But it is not likely that people will use the energy from ocean currents because it would be too expensive.

Riding on the waves

There are two ways to catch the energy of waves. One way is to build power stations on land, near rocky cliffs. The power station uses the energy of the waves as they hit the rocks. But there is a problem with this kind of power station. If there is a bad storm, the waves can be very strong and damage the power station. For example, one wave power station in Norway was destroyed by a huge wave in a bad storm.

There is another way to catch wave energy. Scientists have invented wave machines that float on the surface of the sea. They ride up and down on the waves. One of these machines is called the CLAM system (see above). The CLAM system was invented in Britain. In the future, this system could provide enough energy to make a lot of electricity very cheaply.

The energy of falling water

Water flows down hills because of the pull of the Earth's gravity. People can use the energy from the movement of the water to drive machines, or to make electricity. One way to use the energy of water is to build a waterwheel. As the water rushes downhill it turns the wheel. The movement of the wheel can be used to power a turbine. In 1881, a waterwheel was used to make electricity for the first time. Today, using the energy from falling water is called **hydro-electric power (HEP)**.

Flooding the land

People have built huge **dams** to produce hydro-electric power. A dam is a barrier that holds back water in a lake or a river. The water flows out through special gates in the dam.

◀ This HEP station is in the mountains of Snowdonia National Park in North Wales. It uses a small amount of water falling from a high dam to turn the turbines in the turbine house at the bottom.

There are two ways to get enough energy from the water held back by a dam. In mountains, small amounts of water rush very quickly from a great height. The energy of the fast-moving water powers the turbines. In other places, large amounts of water flow out of the gates in the dam and fall a short distance. The energy from the movement of so much water powers the turbines.

Hydro-electric power is a renewable source of energy, and it does not pollute the atmosphere. But there are problems with HEP. Dams are very expensive to build, and the water held back by a dam often

◄ This old waterwheel used to drive machinery in a mine.

► Inside a large turbine house at a dam in Russia. Water flows out very quickly after passing through the turbines (inset).

floods a lot of land. Sometimes, when a new dam is built, people have to move out of their villages because the water of the new lake will flood their homes and land.

More problems

Dams cause other problems apart from flooding land. The water held back by the dam forms a new lake. The water in the lake does not move very much, and it becomes stagnant. Stagnant water is unhealthy and often dirty. Fish cannot live in unhealthy water, so rivers with dams often do not have many fish or other animals.

Some rivers carry a lot of mud and sand as they flow along. The River Nile in Egypt used to flood farmland next to the river every year, spreading mud and sand on the fields. When the water of the river went down after the floods, the mud and sand was left on the fields. This mud and sand was a good **fertilizer** and helped Egyptian farmers to grow their crops. In the 1960s, engineers built the Aswan High Dam across the River Nile to

◀ A small HEP station in Sri Lanka. This small station makes enough electricity for one village.

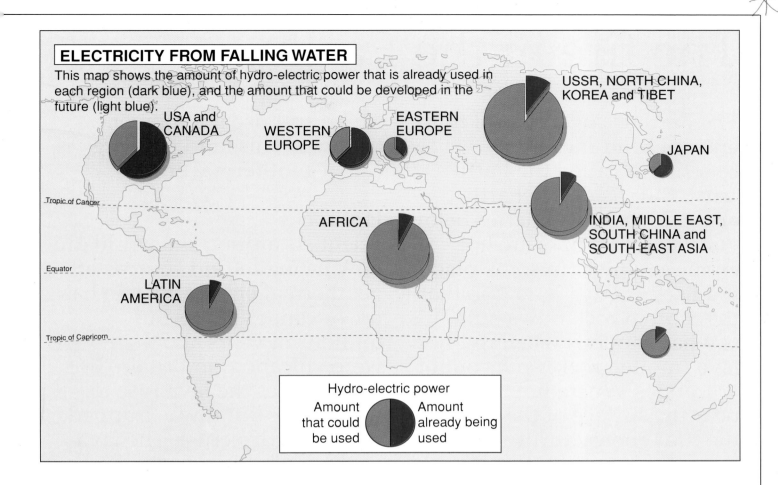

ELECTRICITY FROM FALLING WATER

This map shows the amount of hydro-electric power that is already used in each region (dark blue), and the amount that could be developed in the future (light blue).

USA and CANADA

WESTERN EUROPE

EASTERN EUROPE

USSR, NORTH CHINA, KOREA and TIBET

JAPAN

Tropic of Cancer

AFRICA

INDIA, MIDDLE EAST, SOUTH CHINA and SOUTH-EAST ASIA

Equator

LATIN AMERICA

Tropic of Capricorn

Hydro-electric power

Amount that could be used

Amount already being used

provide hydro-electric power. The dam stops the river flooding. So farmers now have to buy fertilizers to put on their land. The mud and sand in the river is collecting behind the dam. One day, the mud and sand will probably stop the dam working as a hydro-electric power station.

The future for HEP
In spite of all the problems, hydro-electric power is a clean and renewable source of energy. One of the answers to the problems caused by large dams is to build small HEP stations. These small stations use the energy from a stream or waterfall. They only make enough electricity for a small village or town. But they are already successful in many parts of the world.

hydro-electric power — electricity made by turbines that are turned by the force of falling water.
dam — a barrier that holds back water.
fertilizer — substance that helps plants to grow.

The fire in the Earth

The Earth is covered by a thin layer of solid rock. This covering is called the Earth's crust. But inside the Earth, about 20 to 30 kilometres below your feet, the rocks are hot enough to melt. In some places the Earth's crust is so thin that the **molten** rock flows out. This is what happens when red-hot lava (molten rock) spills out of volcanoes. Where hot rocks lie near the surface of the Earth, the heat energy turns water in the rocks to steam. The steam rushes out of cracks in the ground. People can collect this steam and use it to power turbines. This kind of energy is called **geothermal energy**.

Hot rocks

Scientists around the world are now trying to find sources of geothermal energy. Britain has no volcanoes today, but millions of years ago there were volcanoes in Britain. In some places the lava rose to the surface but it was trapped by the Earth's crust. The lava cooled down and formed a rock called granite.

Some of the chemicals in

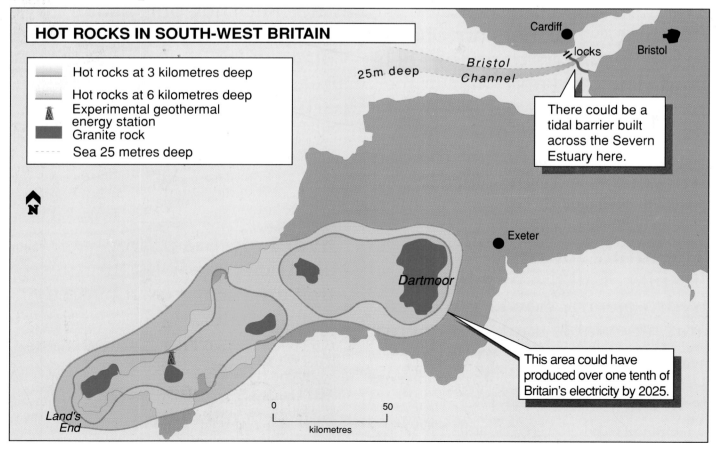

HOT ROCKS IN SOUTH-WEST BRITAIN

- Hot rocks at 3 kilometres deep
- Hot rocks at 6 kilometres deep
- Experimental geothermal energy station
- Granite rock
- Sea 25 metres deep

Cardiff

locks

Bristol

Bristol Channel

25m deep

There could be a tidal barrier built across the Severn Estuary here.

Exeter

Dartmoor

This area could have produced over one tenth of Britain's electricity by 2025.

Land's End

0   50

kilometres

32

granite give off heat energy. They do this because they are radioactive (see page 23). Underground, the heat in the granite is trapped and gradually the rocks get hotter and hotter. In order to use this heat energy, engineers have drilled deep holes into the rocks. They pump water down these holes. The water turns to steam, and the steam is used to drive turbines.

There is only one problem with this source of energy. It is not renewable. Once people have used the heat stored in the rocks, they will cool down. It will take millions of years for the rocks to build up their heat again. This is one reason why the British government has now decided not to develop the scheme.

Hot water

Paris   English Channel   Southampton

In southern England and northern France, engineers have found a huge amount of hot water in underground rocks. This water is hot enough to heat buildings. In Southampton and Paris many buildings already use this hot water. Like the granite rock in Cornwall (see map, page 32), this hot water is not renewable (except over millions of years). But there is enough water to last for many years.

Trapping volcanic heat

In places where there are volcanoes there is also a lot of geothermal energy. In these places this energy is renewable. Many countries around the world have renewable geothermal energy. They include New Zealand, Mexico, the USA, Italy, Japan and Iceland.

Iceland is an island in the north Atlantic Ocean. The island lies across a huge crack in the Earth's crust. In many places in Iceland, molten rock lies just below the surface of the Earth. The people of Iceland use the heat energy from the molten rock to heat

▲ In Iceland, people use geothermal energy to make electricity. The heat from the rocks just underneath the ground is used to turn water into steam. You can see the steam in this picture.

▼ Swimming in hot water in Iceland. The water in this outdoor pool is heated by geothermal energy.

▲ There are hot rocks under this landscape in northern Britain. They could provide geothermal energy in the future.

their homes. They also use the heat to turn water into steam. They use the steam to drive turbines to make electricity.

There is so much geothermal heat in Iceland that it is very cheap to make electricity. Some people have plans to sell electricity made in Iceland to other countries. An underwater cable could take electricity from Iceland, across the North Sea, to Scotland. This could be a cheap and clean way to make electricity for Britain.

molten — something that is molten is so hot that it has become a sticky liquid.
geothermal energy — heat energy that comes from hot rocks in the Earth's crust.

Saving energy

In this book, we have looked at the four main kinds of fuel used to make electricity. These fuels are coal, oil, gas and uranium. We have also looked at many kinds of renewable energy sources. But if people used less energy, this would cut down on the amount of fuel needed. When people try to find ways to save energy, it is called energy conservation.

There are many ways to save energy. People can build houses with thick walls to help keep heat in. This means that less energy is needed to heat the house. Another way to keep heat in is to put layers of thick material, called insulation, in the loft of a house. The insulation stops heat escaping through the roof. Two layers of glass on

▼ Flats in a wall. These flats are in Newcastle, England. They are built along a wall (white in the picture). The flats face south to get lots of heat energy from the sun. The flats have lots of insulation to keep heat energy in.

windows, called double glazing, also helps with energy conservation.

If houses or flats face south, they get lots of sunshine. This means that people can use the heat energy from the sun to keep their homes warm. The heat energy from the sun is free, and it is renewable.

◄ People have put extra insulation on the walls of these old flats. The insulation will help to save energy.

▼ This is a bank in Amsterdam, Holland. The designer of the building planned it carefully for energy conservation. The walls, floors and windows are all insulated. Heat energy from the sun warms the inside of the building.

The house in this picture faces south. The glass windows in the middle of the house let in lots of heat energy from the Sun. This heat energy helps to keep the house warm. In another house, there are large windows on the south side (bottom inset) to keep heat in, and small windows on the north side (top inset) to keep out the cold.

Where to save power

In the developed countries of the world, people use lots of energy in buildings and in vehicles. In Britain, nearly half of all the energy used every year goes to heat and light buildings. Cars and other vehicles use about another third of the energy.

Many poorer, developing countries have lots of sunshine. This means that people do not use much energy for

▼ A new kind of lamp uses much less energy than ordinary light bulbs. These new lamps also last longer than light bulbs.

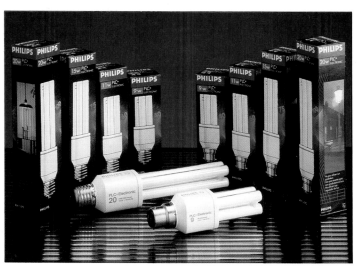

▲ The hood over the cooker in this kitchen takes warm air from the kitchen and uses the heat to warm up fresh air for the rest of the house.

◄ Solar panels collect heat energy from the Sun. The heat energy is used to heat up water.

▲ Special thick insulation in walls and roofs helps to keep heat in, and cold out of a building.

◄ Double glazing helps to save heat energy in a building.

heating their homes. In these countries people use most energy for cooking and for vehicles.

All over the world, people are trying to find ways to save energy, especially in buildings and in vehicles.

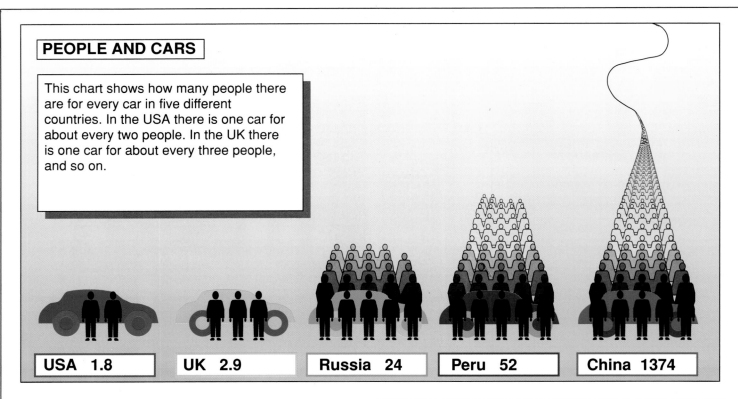

PEOPLE AND CARS

This chart shows how many people there are for every car in five different countries. In the USA there is one car for about every two people. In the UK there is one car for about every three people, and so on.

| USA 1.8 | UK 2.9 | Russia 24 | Peru 52 | China 1374 |

▶ This car runs on electricity. The electricity is stored in a battery. Every 240 kilometres the battery runs out of electricity. Then the driver must fill the battery up again. This is called recharging the battery.

◀ This electric car is recharging its battery at a public electric point in Los Angeles, USA.

Saving fuel in vehicles

In the USA, there is one car for every two people. In China there is only one car for every 1350 people. This is because most people in China travel by bicycle. If everyone in the world owned a car, we would soon run out of fuel. The exhaust gases from the cars would completely damage the atmosphere. To stop this happening, cars of the future will be very different.

Some cars already use new fuels such as electricity, alcohol fuel, and solar power. These fuels are renewable, and they do not pollute the atmosphere. But it will be a long time before everyone is using these new, clean fuels. In the meantime, the best way to save fuel is to drive more slowly. In the USA, people are not allowed to drive as fast as people in Europe, to save energy. There are also fewer accidents because of this. Will other countries follow the good example of the USA?

Energy is needed to make metals and other materials. Recycling these materials saves energy. In the USA, more than 20,000 cars are sent for recycling every day.

Waste materials can be burned to give heat for buildings.

Recycling

Another way to save energy is by recycling. Recycling means using something again instead of throwing it away. When cars get very old, people can recycle the metal parts and other materials to make new cars. Glass and paper can be recycled too.

Every day we throw rubbish into the rubbish bin. Rubbish collectors pick up the rubbish and dump it on a rubbish tip. But we could take out the useful materials before we threw away our rubbish. Then the rest of the rubbish could be burned to make heat energy. In Germany, some towns already use the heat energy from burning rubbish to warm buildings. This is an efficient way both to use rubbish, and to make energy.

Conclusion: less is more

Another example of 'less is more'. Native people who live in the rainforest make only small clearings (left) to grow their food. They move to new clearings after a few years so the forest can grow again. People from developed countries often cut down large areas (right) of forest for mining, farming and so on, and do not plant new trees.

One of the first lessons that an engineer learns about machines is 'less is more'. The machine must do more work and use less energy. For example, engineers are designing cars that can travel longer distances using less fuel than before. This means that new cars are more efficient than older ones.

Many other modern machines are also much more efficient than they were only a few years ago. Televisions, radios and computers all work more efficiently and use less energy.

You have seen in this book that people are looking at many new sources of clean, renewable energy for the future. We must make sure that we use this energy as efficiently as possible, so that we do not waste it. This means that making machines work efficiently will become even more important in the future.

Glossary

crust - the Earth's crust is the thin shell of solid rock that covers our planet.

current - a strong movement of water in a sea or ocean, or in a river.

dam - a barrier that holds back water.

desalination plant - the place where salt is removed from salty water to make it pure and suitable for drinking.

drought - a long period with little or no rain.

ecology - the science that studies how plants and animals are connected with their environment.

environment - the world around us.

fertilizer - a substance that people put on soil to help plants to grow.

fossil fuels - coal, oil and gas are fossil fuels. They come from the remains of plants and animals trapped in rocks for thousands of years.

geothermal energy - heat energy that comes from hot rocks in the Earth's crust.

gravity - the force that makes things fall towards Earth when they are dropped.

hydro-electric power (HEP) - electricity made by turbines that are turned by the force of falling water.

molten - something that is molten is so hot that it has become a sticky liquid

osmosis - the movement of a liquid with high energy (eg fresh water) to a liquid with lower energy (eg salt water) through a screen with very tiny holes.

pollute - make dirty.

power station - a place where fuel is burned to make electricity.

solar - powered by energy from the Sun.

tide - the regular change in level of the sea along seashores.

Index